# WATER

by Michèle Dufresne

Pioneer Valley Educational Press, Inc.

Water is very important.

People and animals need water for drinking.

Plants need water to grow.

Water can be a **liquid**, a **solid**, or a **gas**.

When water is a liquid, you can drink it.
You can swim in it, too.

When water is a solid,
it is ice, snow, or frost.

When water is a solid,
you can ski on it.

Water freezes to form ice
at zero degrees **Celsius**.

Liquid water can **evaporate** to become a gas.

Water is in the air you breathe, but you cannot see it or smell it or taste it.

*The liquid water droplets on this window are slowly evaporating, becoming gas.*

Water covers a large part of the earth.

Most of the earth's water is in the oceans.

In some places,
the earth's water is ice.

At the North Pole and the South Pole,
there are large areas
of ice called ice caps.

Water has many uses.
People use water for cooking, cleaning, and washing.

Water is sometimes used to make electricity. Moving water can make **turbines** go around and around. The motion of the turbines makes electricity.

There are places where there isn't enough water. In some places, water has become **polluted**.

Water is very important for people, animals, and plants. We need to take care of our water supply and not waste it.

# GLOSSARY

**Celsius**: a measure of temperature

**evaporate**: to change into a gas

**gas**: a substance like air that is not a solid or a liquid

**liquid**: a substance that easily flows and takes the shape of its container

**polluted**: made unclean or unhealthy

**solid**: a substance with a definite shape and size

**turbine**: a machine used to make electricity

# INDEX

Celsius 6
evaporate 8
gas 4, 8, 9
liquid 4, 8
polluted 14
solid 4, 6, 7
turbine 13